Trace These Cookies to Use as You Read the Book

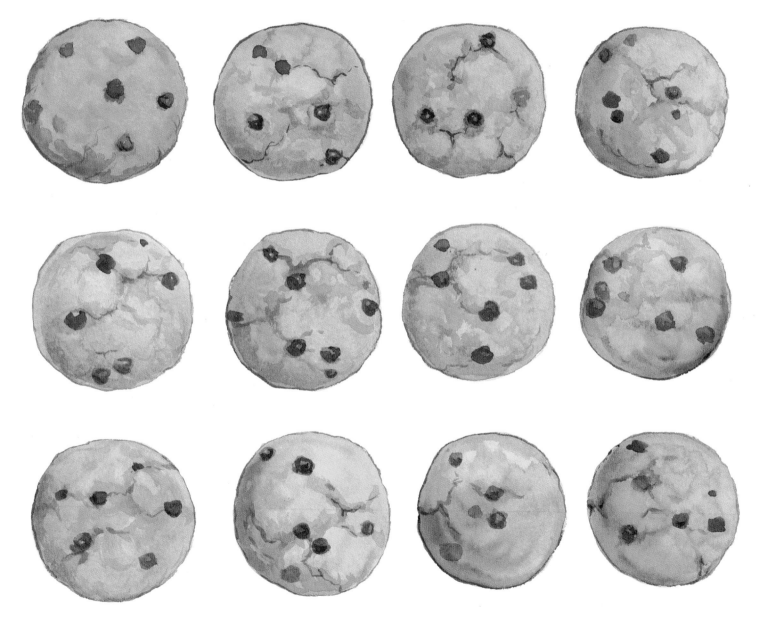

In a cozy kitchen, somebody's baking.
What is Little Red Riding Hood making?
Chocolate-chip cookies for Grandma's tea.
Will there be enough? Wait and see!

Into the garden with the sky so blue
Steps Red Riding Hood as she notices you.
"We're off to Grandma's for a special visit.
Do you have your basket with 12 cookies in it?"

Down in the woods, who can we see?
It's Old Mother Hubbard under a tree.
"Red Riding Hood, my cupboard is bare.
Do you have any cookies to spare?"

Turn the wheel
and choose!

The poor old woman who lives in a shoe
Stops where the path divides in two.
"I have so many children I don't know what to do!
Can you share a cookie or two?"

Turn the wheel
and choose!

At the foot of the hill by the babbling brook
Sits Little Miss Muffet with a hungry look.
"I've just spilled my curds and whey.
What am I going to eat today?"

Turn the wheel and choose!

Over in the meadow where the blackbirds sing
Lies an empty pie and a hungry king.
"I can't eat my pie, as you can see.
Do you have any food for me?"

Turn the wheel
and choose!

Under the bridge where the river runs fast,
Hides a grouchy old troll who won't let her past.
"I'm waiting for those goats to come back,
But maybe you've got a tastier snack?"

Turn the wheel
and choose!

From behind a tree, looking so thin,
Leans a wiry wolf with a hungry grin.
"I was thinking of having your grandma for tea,
Or would you rather give those cookies to me?"

Turn the wheel
and choose!

In Grandma's cottage, everyone's waiting.
But what are they all celebrating?
It's Red Riding Hood's birthday surprise.
Is everyone there? Look—use your eyes!

Math Games

Activities and math conversations help children to see that math can be fun. Here are some more games that adults and children can enjoy together.

AS YOU READ THE STORY

Before the child counts the cookies left in the basket, ask:
How many do you think you will find?

Reset all the wheels to different number combinations, and ask:

How does giving away a different number of cookies change things?

AFTER READING THE STORY

Under the Basket

Turn the basket over and place 5 cookies underneath it. Slide 1 cookie out, without letting the child see those that remain. Show the cookie to the child and ask:

How many do you think are still under the basket?

Lift the basket and count together to check.

Place the 5 cookies under the basket again. This time slide 2 cookies out and ask how many remain. Then repeat, starting with a different number of cookies.

Cookies for the Family

Place 10 cookies in the basket and say:

Let's pretend each person in your family wants to eat one of these cookies!

Ask the child to remove the correct number, one at a time, and to say who will eat each cookie. With each removal, ask the child how many cookies are left in the container.

Baking Cookies

Bake some cookies together. Arrange the plain dough on the baking tray, then decorate the cookies with 1 to 10 chocolate chips, counting them out each time. When the cookies are ready, ask how many you are placing in the cookie jar. Each time cookies are taken out of the jar, ask how many are left.